# ORGANIC EVOLUTION

# ORGANIC EVOLUTION

## OUTSTANDING DIFFICULTIES
### AND
## POSSIBLE EXPLANATIONS

BY

### Major LEONARD DARWIN
HON. Sc.D. CANTAB.

CAMBRIDGE
AT THE UNIVERSITY PRESS
1921

# CAMBRIDGE
## UNIVERSITY PRESS

University Printing House, Cambridge CB2 8BS, United Kingdom

Cambridge University Press is part of the University of Cambridge.

It furthers the University's mission by disseminating knowledge in the pursuit of education, learning and research at the highest international levels of excellence.

www.cambridge.org
Information on this title: www.cambridge.org/9781316633465

© Cambridge University Press 1921

First published 1921
First paperback edition 2016

A catalogue record for this publication is available from the British Library

ISBN 978-1-316-63346-5 Paperback

# PREFACE

THOUGH I have never received any systematic instruction in biology, yet I have heard problems connected with organic evolution discussed ever since I can remember anything. Moreover I have had to consider certain racial questions with great care in connection with Eugenics. The difficulties which the Eugenist has to face are, however, in large measure moral and practical rather than scientific and theoretical; for no biologist known to me doubts that great changes could be made in the human race by weeding out bad stock and by making the higher types multiply more rapidly. It is folly to dream about evolving superman from man in the future; but directly we begin to speculate about how new species came to make their appearance on earth in the remote past we enter on debatable ground differing in many respects from that which must be traversed by those who are seeking to improve the inborn qualities of future generations. In the course of my studies of course I learnt that my father's theories had been subject to severe criticism, especially in recent years; but as regards the probability of making his whole work fall to pieces, these criticisms, so I judged, might be compared to the removal of a dozen bricks from a well built house. Knowing that I might be led away by prejudice in forming this opinion, I searched for some book where none of the objections to accepting Darwin's teachings were either shirked or exaggerated and where an honest attempt was made to fill the gaps created by temperate criticism; but I searched in vain. Though it was straying somewhat outside my proper domain, this led me a few years ago to begin writing and re-writing some brief notes on the points in regard to which it seemed no longer possible implicitly to rely on the *Origin of Species* as a biological guide; and these notes I am now venturing to publish, not so much in the hope that my suggestions, in so far as they are novel, will be endorsed by experts, but rather with the desire to induce some competent biologist to write a book suitable for the general reader in which recent changes of opinion in regard to organic evolution are clearly discussed and wisely criticised—such a book as would prove, as I believe, that with a few exceptions, unimportant as regards broad final results, the views set forth in the *Origin of Species* still hold the field.

When discussing this pamphlet with a scientific friend I was advised to support my arguments by every single fact I could lay my hands on. After careful consideration I decided, however, to adopt the opposite policy and to mention nothing which was not necessary in order to illustrate the views advocated. The amateur who enters the scientific arena can do but little harm and will save himself from heavy blows if he confines his efforts to the suggestion of explanations of unsolved problems. The final decision on all questions connected with organic evolution must rest with students of natural science and genetic experimentalists. I have stated nothing which is contrary to such facts as are known to me; whilst the advocacy in great detail of theories held by me to be true would require more knowledge than I possess in order to be beneficial rather than harmful.

I am glad to take this opportunity of thanking Prof. E. W. MacBride, Mr R. A. Fisher and Mr M. A. Carr-Saunders for help so willingly given in various ways.

L. D.

*August,* 1921.

# CONTENTS

# ORGANIC EVOLUTION

## OUTSTANDING DIFFICULTIES
### AND
## POSSIBLE EXPLANATIONS

### (1) The selection of infrequent mutations and the inheritance of acquired characters could not alone account for evolution.

THE theories of evolution associated with the name of Charles Darwin have been keenly criticised in recent years, and in some respects these attacks are sure to produce permanent results. This must be in a measure the fate of all scientific enquirers who break new ground, and that it would be so in his case was doubtless anticipated by the author of the *Origin of Species*. The criticisms of Darwin's works have, however, not infrequently been unjust, an injustice which has sometimes arisen from ignorance of the aims which he had in view. His objects were to show in the first place "that species had not been separately created, and secondly, that natural selection had been the chief agent of change, though largely aided by the inherited effect of habit, and slightly by the direct action of the surrounding conditions[1]." The first of these objects, that is, the establishment of a belief in descent with modification, was always held by my father to be the more important of the two; for I once heard him say, if a recollection of about fifty years' standing may be trusted, that "after

---

[1] *Descent of Man.* 2nd Edition, Vol. I. p. 92.

all, evolution is the great thing, not natural selection." The public do not now realise how few persons admitted the truth of this "great thing" when he began his enquiries; and as to scientific men, they now look on organic evolution as being such a well established fact that their attention is directed only to questions concerning those methods by which evolution may have been brought about which are still disputable. In consequence they sometimes forget that Darwin was entirely successful in what he regarded as his main object.

If we turn to the secondary objects which the author of the *Origin of Species* had in view, namely the discovery of the ways in which descent with modification has been slowly brought about in the past, we see that since his day great changes have taken place in the generally accepted opinions concerning the inherited effects of habit. Weismann succeeded in nearly destroying for a time all belief in the racial effects of use and disuse, whilst now the pendulum is swinging somewhat back again towards the acceptance of the inheritance of acquired characters as a factor in evolution. It is not my intention now to touch on this controversy, or to discuss to what extent, great or small, use and disuse have been the direct causes of evolutionary changes, except merely to state my reasons for holding that in respect to many or most structures the inheritance of acquired characters must at all events have been aided or controlled by natural selection in order to have produced the organic forms now in existence.

In the first place the amount of use to which in bygone times different structures of the same organism were being put cannot in many instances have tallied with the need for their growth in order to have produced the innumerable complex adaptations now found in nature. To illustrate this point, the development of the maternal instincts may be cited as a single example. Amongst many animals we find that the mother's affection for her young forces her to risk her life freely for their protection, whilst it does not so completely overcome her sensations of fear as to prevent her from abandoning her progeny when the danger becomes extreme. This balance of instincts seems to make for the survival of the species, and it may, therefore, have been produced by natural selection. But there seems no reason why a state of things tending to promote survival should have resulted from the inherited effects of use; for the relative strength and duration of the sensations of fear and affection do not necessarily bear any relationship to the need for the development of these qualities in such a way as to assist in the struggle for existence.

A more important difficulty in the way of those who seek to attribute evolutionary effects to the *unaided* influence of the inheritance of acquired characters may be illustrated by reference to the growth of muscles through the stimulus of exercise. If the growth of a muscle due to exercise in any one generation merely resulted in the *same* amount of growth following on the average the *same* amount of exercise in the *next* generation, this would be the result which we should anticipate judging by the ordinary laws of inheritance; and we should find in this fact no explanation of the process by which the muscle had been evolved so as to have attained its present dimensions and, therefore, no clue to the process of evolution. To afford such a clue, must not the inheritance of acquired characters mean in this instance that the growth of a muscle through exercise in one generation would result in that muscle being *more* developed in the coming generations after having received the *same* amount of stimulus from exercise? Muscles are, however, used in every generation, and if this were the only relationship which existed between exercise or use and racial development, would it not follow that in every animal each muscle would always be bigger than was that same muscle in the preceding generation? If there were no other agency at work but the inheritance of acquired characters, would not the inevitable evolutionary result be an indefinite increase in the size of every muscle? In many cases obviously this would have produced extremely harmful results of a kind that are in fact seldom found in nature; and it must, therefore, be assumed that some other agency has always been at work preventing these evil consequences from arising. Can any other brake but natural selection be suggested as that by which this tendency for use[1] to cause an indefinite racial increase in the size of all the structures used has been brought to a standstill?

But even if natural selection be accepted as having been an efficient co-operating agency in preventing overgrowths in the evolutionary process, our difficulties are not thus all cleared out of our path. Structures have often grown less and less as the generations succeeded each other until they quite disappeared ; and, to account for this disappearance,

---

[1] For the inheritance of acquired characters to have produced very wide evolutionary effects, must not 'use' include the growth or development of all organic structures?

'disuse' has often been suggested as having been the cause. Even if it be admitted that without selection disuse may form the basis of a rational explanation—and possibly the only rational explanation—of the disappearance of some *useless* structures, we yet have to account for innumerable instances in the evolutionary process of *useful* structures having merely become diminished in size without having vanished. Those who in such cases also rely on 'disuse' as having been the cause can hardly have considered what meaning they wish to be attached to the words they use; for the structures thus held to have been diminished in size by 'disuse' must in many cases have been in constant 'use' in every generation. But if we should decide to adopt unusual meanings for our words, making the 'disuse' of a muscle merely imply that it is not being as much used, and 'use' imply that it is being more used than was that same muscle in the *previous generation*, then we shall find that use and disuse as thus defined may very likely be properly relied on when laying the foundations of a theory of organic evolution which is at all events logical. To illustrate this point allusion may be made to the suggestion that internal secretions form the connecting link between the structure and the germ-plasm by means of which use and disuse are made to produce hereditary effects. If the average growth of a muscle in a given generation resulted in the production of a certain quantity of the appropriate secretion, and if this particular *quantum* of the secretion were such as to produce no effect on the germ-plasm, then in an evolutionary sense the organism would be in a state of equilibrium as far as the development of that particular muscle was concerned. But if a greater use of the muscle, besides increasing its size, were to make it produce a greater amount of the secretion, and if the *additional amount* of the secretion produced above the *quantum* were so to affect the germ-plasm as to make the muscle become larger in the next generation after the same amount of use, then this acquired character would be inherited. In the same way we can conceive that a reduction in the size of the muscle through lessened use might also be a heritable character in consequence of the secretion being produced to an amount below the *quantum*. It would, however, still have to be admitted that an increase or a decrease in use above or below the original amount of use would lead to an indefinite increase or decrease in the size of the muscle unless this tendency were checked by some such agency as natural selection. In fact there must exist some method of restoring racial equilibrium when an organism has again come to be well suited to its surroundings, a restoration which would be accomplished if the different secretions came to be produced only in such quantities as to have no effect on the germ-plasm. May not this result be brought about by natural selection in the following manner? Let it be assumed that the size of the *quantum* of the secretion which does not affect the germ-plasm is a variable quality or differs amongst the different individuals of the species[1]. Taking the case of a muscle which had been increased in size by the inherited effects of use up to a point where a further increase would on the whole be harmful, natural selection would then begin to come into play by eliminating not only those individuals which would in any case possess exception-

---

[1] The sensitiveness of the germ-plasm might also be the variable.

ally large muscles, but also those in which the production of the secretion was exceptionally large. Would not this process go on until, by the survival of individuals both well suited to their environment and producing such a *quantum* of the secretion as would not affect the germ-plasm, a state of equilibrium had again been established?

Thus we see that by assuming the efficacy of either secretions, pangenesis, or any other connecting link between use and disuse and racial development, a system of inheritance of acquired characters does become a logical possibility. Judging by the way in which useless characters have disappeared, and considering the necessity for some mechanism resulting in a restoration to a state of equilibrium, evolution as the result of use and disuse seems likely to have been an exceedingly slow process; a conclusion which in no way tells against a belief in this method of producing racial changes. It has, however, been claimed that the inheritance of acquired characters is the simplest explanation of evolution; but if it necessitates the admission of the efficacy of natural selection, and if natural selection has to be brought into play in this complex manner, this claim becomes at least very doubtful.

In the foregoing discussion on the inheritance of acquired characters I have tacitly assumed the absence of any vitalistic principle or continuous modification of the action of the forces of nature. To discuss the validity of making any such assumption would open out a subject too wide here to be dealt with; and I will only say that what I have assumed does not necessarily involve the belief that science alone can solve the riddle of the universe. As to the direct effects of environment, that is as to mutations directly caused by the action of external conditions on the germ-plasm, this is a factor on which Neo-Lamarckians have to rely to a considerable extent; and here it is even more evident that they have to choose between placing reliance on selection as a necessary co-operating agency and trusting to some vitalistic principle. If mutations directly due to environment were only to make their appearance in directions making for survival, it would be necessary to hold that their production was guided by some unknown external agency or by some *sentiment interieur*, to use Lamarck's phrase. But to rely on any unexplained law limiting mutations due to environment to such as do make for survival is almost to abandon a belief in organic evolution; for the essential feature of such a belief is to hold that descent with modification has been brought about by natural causes all of which might still be operative. Those who believe that evolution is now explicable must, therefore, maintain that the nature of these environmental mutations has no relationship to the survival of the species, and that it is only natural selection which makes them play a useful part in the evolutionary process. In fact it seems that the chief function of use, disuse, and the direct effect of environment has been to create mutations on which selection was able to act effectively. The main point here insisted on is, however, that for the foregoing reasons, together with others which I am incompetent adequately to discuss—such, for example, as those based on the wonderful instincts of worker bees—we must hold that the inheritance of acquired characters and mutations due to environment cannot alone account for the appearance of all the existing forms of organic life. Rule out selection and evolution becomes inexplicable.

We must, therefore, next consider whether selection can in truth do all that is necessary to make up a complete theory of organic evolution. The controversies concerning the part played by natural selection have often turned on whether the mutations which are necessary to serve as a basis on which it could act have been large and infrequent (and to these the term mutation has often been exclusively applied), or small and frequent (which have been inaptly described as continuous variations). Adopting the word 'mutations' to cover all original modifications in the germ-plasm, I am not here concerned to enquire what part has been played by large mutations in the evolutionary process, but merely to indicate as briefly as possible why they cannot supply all that is needed in the way of explanation.

The difficulty of always relying on large mutations, which never do occur frequently, becomes most readily apparent when studying those cases of adaptation which must have necessitated modifications of a suitable kind having taken place nearly simultaneously in several different parts of the same organism; the classical instance being the increase in the weight of the stag's antlers and the necessary increase in strength and size of the neck and shoulders to bear the greater burden. It has not, I think, been as a rule sufficiently clearly realised how extraordinarily seldom all the necessary mutations would occur at the same time in the same individual. Imagine, for the purposes of illustration, a mutation which occurs on the average once in every thousand organisms; that is such a mutation that, if it occurred in man, would show itself in about fifty thousand inhabitants of the United Kingdom. Then it can be demonstrated that, if dealing with mutations which are in no degree correlated, it would be as likely as not that we should have to wait for over ten thousand generations for four of these mutations to appear in these islands simultaneously in any one individual. That all four of these large mutations would, when they did appear, all be of about the right relative magnitude to produce proper adaptation, would be very improbable. And even granted that such a prodigy of adaptation would appear at extraordinarily rare intervals, the newly appearing Mendelian factors would become widely scattered amongst different organisms after a few generations of random mating; the proper adaptation would no longer exist ; and if the changes of character were in themselves harmful when standing alone, these mutations would be eliminated by selection during the immense period of time which would elapse before they would all be again united in one individual. Whatever it may be which produces a change in one structure is no doubt likely to produce changes in other structures also. But unless we can assume that all the changes due to one mutation will be so related to each other as to produce the needed adaptations (as the result of some *sentiment interieur?*) the arguments against the efficacy of large mutations which are based on probabilities, though changed in form, yet are not invalidated by the fact that several parts of the organism are often simultaneously affected; for the greater the number of structures suitably affected by one mutation, the rarer would be the occurrence of such a mutation. It follows that if trusting entirely to the effects of large and infrequent mutations, every adaptation must have been the result either of a number of mutations each one of which would be beneficial if occurring alone, or of a mutation in a single Mendelian factor, or of mutations in completely linked groups of factors. Will any student of Mendelian heredity hold that all adaptations could have been brought about in one of these ways?

As to small mutations, they no doubt would often be beneficial when large mutations in the same direction would be harmful. But mutations when both small and infrequent would, if injurious when appearing alone, be ineffective in producing useful changes in a number of uncorrelated characters; for their simultaneous appearance in the same individual would be rare and their coincidence would not be maintained in future generations. It is true that, as compared with large mutations, small mutations would not be as readily weeded out by selection; but this would be equally true of mutations unsuitable to adaptations as of those which were suitable. In the higher animals, at all events, a change in nearly every important structure would necessitate an otherwise harmful change in some other structure in order to be on the whole beneficial; and a theory of evolution must include a method of accounting for adaptations in these circumstances. Large mutations never occur frequently, and as we have seen that infrequent mutations, whether large or small, cannot fill all the parts needed, our hope of finding a complete explanation for the process of evolution seems at present to rest on being able to place reliance in some measure, great or small, on the selection of small and frequent mutations[1].

---

[1] The word 'frequent' is here used in relation to the rate of the evolutionary process. The slower that rate—and I believe it to have

## (2) To admit the selection of small and frequent mutations amongst the explanations of evolution demands the solution of several unsolved problems.

The measurements of the characters of organisms belonging to the same species in a state of nature as a rule are found not to be identical, but to be nearly uniformly distributed about a mean measurement; and if in regard to each separate structure these individual differences do form a basis on which natural selection can act, then it will be seen that the above-mentioned difficulties in accounting for the simultaneous modification of a number of different structures would not be felt. The difficulties in question, it was seen, arise because it must be assumed that several independent qualities have often been *simultaneously* modified by natural selection, even when the necessary variations would have been harmful when standing alone; and that, for this to have occurred, the beneficial variations of all these qualities must have been present sufficiently often *in the same individual* at the same time. If as regards each one of these independent qualities there did exist a series of differentiated individuals grouped about a mean form, as is generally the case in nature, then it would follow that half the population would be above the average in regard to any one of these qualities, one quarter in regard to any two of them, one eighth in regard to any three, and so on. There would, therefore, always be under these conditions a number of

been normally exceedingly slow—the longer may have been the intervals in time between these 'frequent' mutations. Moreover the expression 'frequent mutations' here always implies the frequent appearance of mutations of the same kind.

individuals each one of them above the average in regard to any moderate sized group of qualities; and if these individuals were to be selected for survival —individuals who might be regarded as all simultaneously advancing in the right direction—it would appear at first sight that the necessary adaptations could thus certainly be evolved. The rapidity of the process would of course depend on the severity of the selection and on the number of the qualities to be simultaneously dealt with. With regard to man it has been stated that one half of the population of any one generation is descended from one ninth of the population of the preceding generation; and if this be so and if the selection which produced this result had been evenly concentrated on any three qualities, it would follow that more than one half of the population would be descended from parents all conjointly above the average in regard to all these three qualities[1]. Such a complete concentration would rarely occur, but amongst the lower organisms selection acts far more severely than this; and we thus see that with a small group of independent qualities, the necessary coincidence of suitable variations in the same individual presents no formidable difficulty in explaining the evolutionary process. Natural selection would moreover, as it were, move about from one line of descent to another as the generations succeeded each other until a close adjustment between a number of independent qualities had been evolved; and to Darwin it seemed that adaptations of any degree of nicety could thus slowly be produced. Individual differences certainly do exist in nature, but we must enquire whether they indicate the presence

[1] Dr Heron. *Biometrika*, Vol. x. p. 419.

of true mutations, or of variations of such a kind as to enable them to serve as a basis on which natural selection could act. In other words, can we assert that the existence of these differences does afford a proof that organisms can be permanently and progressively modified by selection, and that new types now never seen on earth may be expected constantly to make their appearance under the pressure of competition? These are the questions now to be considered.

Though large and infrequent mutations certainly do occur in nature, it now seems to be held by certain biological experts that we have no incontrovertible proof that small mutations ever do frequently make their appearance. According to their views, all the individual differences observed in a species are due to the rearrangements of the allelomorphs which inevitably occur at each successive generation; and the belief that this is the case is no doubt supported by the fact that the laws of natural inheritance can be well explained on the hypothesis that each character is due to one or more *unaltering* Mendelian allelomorphs. If the allelomorphs are in truth unalterable, then one generation can only differ from the preceding one in consequence of such rearrangements, and selection can only make such racial changes as can thus be produced, the possible results being, therefore, strictly limited. If this be true, it seems to me that Darwin was wrong in holding that the individual differences always noticeable in species form a basis on which natural selection is able to build so as to produce great changes in structure. Under certain conditions considerable changes unquestionably do result from the re-arrangements of the allelomorphs, and it has been urged that all changes in species have been due to crosses with other species having taken place in the past, the necessary variety in the ingredients having thus been produced. On this hypothesis, when a species becomes so isolated or differentiated as to preclude all chances of crossing with other species—as in the case of man—then the evolutionary process could not proceed beyond a limit dependent on the possible effects of further re-arrangements of the allelomorphs; a condition of things which, if it existed, would often have brought evolution to a standstill. Existing organisms of common descent must, moreover, have all sprung from one original zygote; but no logical objection can be raised against the crossing hypothesis on this account, if it be assumed that the original zygote contained all the existing carriers of the hereditary influence in such a group, that is probably all existing genes[1]. If evolution had been such a process of unpacking and redistribution as is thus assumed, then it would follow that all these innumerable genes must have sprung together into existence at one bound, and this seems to me a highly improbable and indeed almost inconceivable hypothesis. This theory had its origin mainly in certain conclusions drawn from experiments on 'pure lines,' conclusions in regard to which doubts have recently been expressed. When plants are continually cross-fertilized, the uniting gametes must become more and more likely to be identical; and as certain investigators believed that in the pure lines thus formed they found no changes were produced by

[1] The argument to be considered later on, which is based on the fortuitous extinction of extreme types, points to the probability of all existing organisms being descended from either one or but few original types.

continued selection, it was assumed that this proved that no mutations were taking place or ever would take place. But even granting the truth of the facts alleged to have been observed, this conclusion cannot be logically deduced from them; for all that could be held to be proved is that, when these identical gametes were thus united, no mutations did take place of such a magnitude as to have been observable in the number of generations covered by the experiments. That crossing can form the basis on which the whole evolutionary process depends seems, therefore, to be neither inherently probable nor a necessary deduction from the facts, and we have no right to assume that genes have been immutable in either number or kind. Crossing may at times have produced rapid evolutionary results; but the existence of one evolutionary process, which can be comparatively easily imitated in experimental research, affords no proof whatever that there do not exist other evolutionary processes the results of which are difficult to detect by the investigator. There is no justification for the assumption that Mendelian factors are never subject to such small changes in successive generations as to have thus far eluded unquestionable detection by the students of genetics. The occasional appearance of large mutations is generally admitted; but may not mutations which are so small that their appearance is readily obliterated by the larger variations due to Mendelian rearrangements also occur? Experimentalists are unquestionably right in demanding that every doubtful theory should be as far as possible tested by experiment; but to hold, as some of them seem to do, that a theory must be altogether rejected until proved to be true by experiment is illogical. If it had not been held to be permissible to adopt provisionally a theory incapable of direct proof or disproof by experiment, the progress of science would have been brought to a standstill in many directions. We are always justified in adopting as a working hypothesis that theory which, whilst necessitating no disproved assumptions, on the whole fits in best with all the facts of the case ; and granted that small mutations *may* frequently take place, and also that neither crossing nor the selection of infrequent mutations nor the effects of use can singly or together constitute a sufficient explanation of evolution, nothing which has been said so far should rule out the selection of small mutations as the most probable hypothesis when seeking to complete the explanation of organic evolution.

There is, however, another and, as it seems to me, a more important criticism which can be brought against the *Origin of Species*, namely, that that work does not fully explain the origin of species, even granting the existence of small and frequent but undetected mutations. The existence of different kinds of animals and plants without intermediate forms, or the segregation of organisms into species, is a fact so familiar to our minds that it often fails to demand the explanation which it ought to receive. The fact of evolution is now almost universally accepted, and, this being granted, it follows that innumerable transitional varieties connecting different species with a common ancestor must have existed in the past, and some explanation ought to be forthcoming as to why these intermediate forms have disappeared, especially as to their disappearance in regions inhabited by both the now clearly distinguishable varieties. " This difficulty for a long time quite confounded " Darwin, and this may perhaps make it less presumptuous to question the ways in which he thought it could "be in large part explained[1]."

Whenever extinct intermediate forms were less well adapted for survival than the forms which have as a fact survived, the elimination of these vanished types is in no way surprising. The extinction of a whole genus or species in the struggle for existence might well leave a wide gap between the remaining forms of life, and it is the existence of gaps between existing varieties which is most difficult to explain. Taking the case mentioned by Mr Bateson as a good example, when plants had been modified by the increasing temperature after the close of a glacial period, and after each modification thus formed had ascended the mountain slopes so as to retain the temperature which was suitable to its needs, in these circumstances we cannot account for the disappearances of the intermediate forms which subsequently took place as being a consequence of their unsuitability to their environments. In such cases we might expect to find the most perfect gradation of forms to have remained in existence, and we have to account for the fact that species very often meet and interlock, and yet that they still generally remain "absolutely distinct from each other in every detail of structure." For such an extinction of intermediate types Darwin suggested several explanations. As regards the effects of selection, his main contention was based on the fact that "forms existing in larger numbers will have a better chance, within a given period, of presenting further favourable variations for natural

[1] *Origin of Species.* Chap. VI. pp. 134, 135.

selection to seize on, than will the rarer forms which exist in lesser numbers." Hence he argued that "the more common forms, in the race for life, will tend to beat and supplant the less common forms, for these will be more slowly modified and improved[1]," and that as groups adapted to different localities and varieties generally are certain seldom to be equal to each other in population, the disappearance of intermediate types will always be taking place. The chances of survival must, however, increase with the number of times the same new form makes its appearance; from which it appears that the frequency with which the mutation appears is as important a condition as the size of the population. Once a beneficial mutation has survived for a few generations, the chances of its extinction become very small; and when this is the case, it matters little whether the surrounding population be large or small. This argument, therefore, only applies with force to the large and infrequent mutations which are not now under consideration; and this reason for anticipating the disappearance of intermediate groups has little weight as far as the effects of frequent mutations are concerned. It is, however, more important for the purposes of my argument to note that, for reasons which will be discussed later on, Darwin had to rely on the unifying effect of inter-breeding on subsequent generations; and if inter-breeding does produce such an effect, then certainly the regular appearance of small mutations would produce the same evolutionary results whatever might be the magnitude of the population[2]; for favourable mutations of equal pro-

[1] *Origin of Species.* Chap. VI. p. 136.
[2] If inter-breeding does make for uniformity of the factors, then the

portionate frequency would always be made to disappear by inter-breeding with the same rapidity amongst large numbers as amongst small.

There are, however, other effects of inter-breeding, assuming it to make for uniformity in future generations, which I am doubtful whether Darwin had in his mind. Taking as typical the case of a number of varieties inhabiting contiguous areas with a limited amount of inter-breeding between them, it may be assumed that each one of them had been modified to some extent by natural selection so as to have made it more suitable to its own surroundings. Now if inter-breeding between any two of these varieties would tend to make them become more like each other, then its effects on each one of them would be to modify it in such a way as to make it more suitable to the conditions obtaining in the area inhabited by the other; in which case the results would obviously often be injurious. But as a given amount of inter-breeding would thus produce less change in the characters of a variety in the coming generations in proportion to the size of the population which it had to affect, it certainly follows that the larger the population the less would be the injury due to inter-breeding. Varieties with large populations would therefore tend to prevail over such as had smaller populations if there were any inter-breeding between them. Moreover if a variety had other varieties on both its flanks, it might thus receive a double injury by inter-breeding, and consequently be more likely to be reduced in

argument also falls to the ground in the case of infrequent mutations; for though such mutations would appear more often in a large population, they would have, as it were, a proportionately greater difficulty in overcoming the swamping effects of inter-breeding.

In other words, chance is continually eliminating some of the extreme types in every species. Here is a common tendency making for uniformity which must be taken into account. It is, however, an effect which will continually become slower and slower, whilst the larger the population the less rapid will be its action; and those who rely on its efficiency must not rule out other processes because they also must have been exceedingly slow[1]. The tendency of natural selection to preserve those types which are best suited to their environments must, moreover, also always be taken into account; with the result that it must be admitted that any variety possessing any slight advantage in relation to its environment would probably not be exterminated by any such fortuitous process. We therefore see that the disappearance of intermediate types, when each type had been adapted to its surroundings, cannot be explained in this way. Then again the case of characters dependent on single factors must be held in view; for abrupt differences in useless structures due to such factors would only be very slowly obliterated and might endure for vast ages in large populations. As regards useful structures, in so far as the elimination of types is a matter of chance, structures dependent

---

[1] These conclusions as to the relative effects on large and small populations are, I am kindly advised by Mr R. A. Fisher, capable of mathematical proof; and, as he suggests, they give a reason for the greater variety existing in large species. Does not this unifying process, moreover, afford an opportunity for testing whether small mutations do take place? In the case of a domestic animal, such as the guinea fowl, which is believed not to be of mixed origin and which is not generally kept in large numbers, identity between the gametes ought by now to have been produced in the absence of mutations. If breeding experiments with guinea fowls should discover the existence of substantial parental correlation coefficients, would not this prove that mutations had occurred?

was in the end fully realised by Darwin; though when writing the *Origin of Species*, according to his own views as expressed at a later date, he did not then take sufficient account of useless structures, this being one of his "greatest oversights." In his *Descent of Man* he suggested that the uniformity of existing species in regard to characters not subject to the action of natural selection is due to an assumed uniformity in the causes, whatever they may be, which have produced "the numberless slight differences between the individuals of each species[1]." But are not those who believe in the efficacy of the selection of individual differences thus asked to rely on an argument which must prove to be most damaging to their own theories? If the causes of all germinal changes affecting individual structures had been so uniform in their action as to have produced absolutely uniform results, then there would have been no slight heritable differences on which selection could have acted. Those who believe that evolution has been in any degree brought about by natural selection acting on individual differences must hold that the causes of these differences, whatever they may be, have always been in action and have always produced a want of uniformity in the organisms belonging to the same species. This necessarily postulated want of uniformity must also apply equally to useless structures, and they also must always have exhibited a certain want of uniformity; for we may assume that the causes at work cannot have distinguished between useless and useful structures. If this be so, what we have to ask is why these same causes should ever have reversed their action and, by putting on a brake, have pre-

[1] *Descent of Man.* Chap. II. p. 93.

vented that ever increasing want of uniformity which we should *a priori* have expected to find in useless structures as the result of any causes of differentiation if they had once been set in operation ? This is the problem, of which possible solutions will be considered later on. Here, however, the point to note is that Darwin accounted for the uniformity of the characters of useless structures in two ways, only one of which has thus far here been mentioned; for he held that this uniformity would naturally follow not only "from the assumed uniformity of the exciting causes," but "likewise from the free inter-crossing of many individuals." If the foregoing considerations force us to abandon the first of these two explanations both as being unsatisfactory in itself and as fatal to the theory of the selection of small mutations, that is if we have to reject the uniformity in the action of the causes of individual differentiation as a method of accounting for the uniformity of useless structures, then, if we follow Darwin at all in this matter, are we not compelled to trust to the effects of inter-breeding as the only remaining explanation ? Are we justified in relying on the unifying effects of inter-breeding, either in regard to useless structures, or as explaining the way in which specific differences have arisen, as suggested in a previous paragraph? That is the question now before us.

## (3) The existence of a system of mutations due to imperfect segregation is suggested as one of the possible explanations.

As to producing uniformity in structure or appearance, the re-arrangements of the allelomorphs through inter-breeding will certainly have that effect when the structures or appearances are dependent on a number of factors; the characteristics of the Mulatto being a familiar instance of the results of this process. Useless structures have often doubtless thus been rendered more uniform. Moreover, as to the elimination through inter-breeding of relatively small varieties, this also would be a consequence of this unifying process; for the smaller the variety, the more likely would it be to be thus made unsuitable to its environment. In a measure, at all events, an explanation is thus forthcoming for the facts above mentioned as needing elucidation. But in the case of characters dependent on one or few factors, no such unification is to be expected as the result of inter-breeding; and, if this were the only explanation of uniformity, we should look not infrequently to see in useless structures marked and abrupt differences between the individuals composing a species.

Another unifying tendency, which is indefinitely progressive in its action, and which has lately been emphasized with great ability by Dr Hagedoorn, may be briefly indicated in the following manner[1]. We have all of us known cases of a "family dying out," or the disappearance of a surname, and on consideration it will be admitted that this must occur in every generation. This obviously implies a continued diminution in the number of male ancestors from whom the population is descended; and, as the same diminution must occur in the case of female ancestors, we see that there must be a progressive diminution in the number of types which together compose the population, or a continued lessening of racial diversity.

[1] *The Relative Value of the Processes causing Evolution.* Hagedoorn. 1921.

In other words, chance is continually eliminating some of the extreme types in every species. Here is a common tendency making for uniformity which must be taken into account. It is, however, an effect which will continually become slower and slower, whilst the larger the population the less rapid will be its action; and those who rely on its efficiency must not rule out other processes because they also must have been exceedingly slow[1]. The tendency of natural selection to preserve those types which are best suited to their environments must, moreover, also always be taken into account; with the result that it must be admitted that any variety possessing any slight advantage in relation to its environment would probably not be exterminated by any such fortuitous process. We therefore see that the disappearance of intermediate types, when each type had been adapted to its surroundings, cannot be explained in this way. Then again the case of characters dependent on single factors must be held in view; for abrupt differences in useless structures due to such factors would only be very slowly obliterated and might endure for vast ages in large populations. As regards useful structures, in so far as the elimination of types is a matter of chance, structures dependent

---

[1] These conclusions as to the relative effects on large and small populations are, I am kindly advised by Mr R. A. Fisher, capable of mathematical proof; and, as he suggests, they give a reason for the greater variety existing in large species. Does not this unifying process, moreover, afford an opportunity for testing whether small mutations do take place? In the case of a domestic animal, such as the guinea fowl, which is believed not to be of mixed origin and which is not generally kept in large numbers, identity between the gametes ought by now to have been produced in the absence of mutations. If breeding experiments with guinea fowls should discover the existence of substantial parental correlation coefficients, would not this prove that mutations had occurred?

on single factors would be as likely to be preserved as any others; and, as long as any differences existed within a species, there would, as far as this method of unification is concerned, be no reason to expect that the distribution would be continuous. Whether these two methods of promoting uniformity are sufficient to account for all the facts needing explanation, I must leave to experts to decide. But as they do not seem to overcome these and other difficulties, I am venturing to suggest for consideration another process which might be always at work tending to create uniformity *in the factors themselves* as well as in characters resulting from those factors.

The hypothesis which I now suggest for consideration may be illustrated by reference to the short and tall sweet peas of Mendelian fame. If it be assumed that, when tall and short sweet peas are crossed, the allelomorph of the tall sweet pea, gives up a minute portion of its tallness to the allelomorph of the short sweet pea—if such a colloquial and unscientific method of expression may be permitted here and elsewhere—then the result would be that the offspring, or rather the pure homozygous descendants of such a union would always be somewhat more nearly equal to each other in stature than were their homozygous progenitors; and by this means a group of freely inter-breeding tall and short sweet peas would be slowly converted into a homogeneous group of medium-sized sweet peas, all breeding true to their kind. Minute mutations are in fact, it is suggested, continually taking place of such a nature as to make the characteristics of the descendants of the original progenitors of any group approach nearer and nearer to the mean of the characteristics of those original progenitors; and these mutations will, therefore, be described as centripetal mutations.

Influences making for uniformity, or centripetal influences are, therefore, always at work, and one or more of these influences will remain in operation as long as there exists any diversity amongst the individuals composing a species. Now whenever mutations do make their appearance at random, they will create a tendency in opposition to the tendencies just described, or a centrifugal tendency; and, if natural selection acting on individual differences has been a factor in evolution, we must assume that mutations have appeared with sufficient frequency to have caused the necessary diversity of hereditary types in spite of all these unifying influences. We must also assume that the centripetal influences, like the hypothetical centripetal mutations, have been unceasingly in operation and thus have prevented these constantly occurring mutations from having caused an ever increasing diversity in the ranks of all species. What is it which causes mutations to take place is a much disputed point. It may be that they are due to various environmental influences. But may it not also be that they are caused or promoted by something which occurs when the gametes are united. Taking the case of the crossing of tall and short sweet peas again for the purpose of illustration, may it not be that the transfers between the allelomorphs above described are not the only ones that are taking place, but that during the union in the zygote minute portions of whatever it is which makes for tallness are also frequently transferred before segregation from the allelomorph resulting in the shorter of the two sweet peas to that producing the taller plant;

both allelomorphs thus being modified so as to make them diverge more from the mean of the two? Taking the case of a group of sweet peas of approximately the same height, the result of this type of mutation if acting alone would be that plants both shorter and taller than any previously existing would begin to make their appearance; the group would become more and more diversified, or in other words, would have a wider range of variation about its mean; and this scattering process would go on generation after generation, though more and more slowly, yet to an indefinite extent. Mutations of this kind will therefore be called centrifugal mutations.

Now may not these two sets of influences, one making for uniformity and the other for diversity, balance each other so as to maintain a constant range of differentiation in all freely inter-breeding groups of organisms? This result might be produced in several ways. For example, the unifying influence due to the fortuitous extinction of extreme types will become slower and slower as it proceeds, and at some definite point it might balance any steady tendency towards diversity due to small and frequent mutations. Or the centrifugal and centripetal mutations might only at some one point produce equal effects in opposite directions. Assuming this to be the case, any nearly uniform group of sweet peas, instead of becoming indefinitely extended from the effect of centrifugal mutations acting alone, would be brought to a position of equilibrium after a certain number of generations by the opposing action of the centripetal mutations; and the group would then for an indefinite number of generations remain normally and uniformly distributed about a mean with a given range of variation[1]. Reverting to the case of two inter-breeding groups of sweet peas, one short and the other tall, it was seen that these groups if only affected by centripetal mutations would be slowly converted into a single group of absolutely uniform plants. If both types of mutations were simultaneously in operation, this coalescence would however be brought to a standstill when a certain range of variation had been reached. In fact if there did exist a suitable relationship between the effectiveness of the two kinds of mutations, then a group of tall and short sweet peas, when continually crossed together, would in time form a group of slightly divergent individuals, *uniformly scattered about a central type* in regard to the measurements of their characteristics; and this would remain the distribution of the group for however long this combined process of centrifugal and centripetal mutations were to go on. The two varieties of tall and short sweet peas would have been slowly converted into a single variety of medium-sized plants.

The foregoing hypothesis may be stated in other words by asserting that the individual differences always found in species are in part due to the counter-play of opposing types of influences and minute mutations, and that as thus formed they would serve as a basis on which natural selection could act. No doubt it has often been held that mutations so small as not to be detected would not have sufficient 'survival value' to form the steps by which great evolutionary results could be produced. But those who argue thus may make the erroneous assumption of holding that each mutation must be separately

[1] In any case the centripetal influence due to the fortuitous diminution of extreme forms would have little effect in experimental research.

established by the action of natural selection before others can be added to it, and in not admitting that considerable numbers of small mutations may be automatically accumulated in one individual and not in another so as to produce a material difference in survival value between them. If the action of these centrifugal and centripetal influences could bring a group of organisms into a condition of stability as regards variation, as is here assumed, the individuals who differed most from each other in regard to any one quality might well differ materially as to their suitability to their environment. Moreover, even if breeding operations with careful *artificial selection*, would take a hundred generations to produce any forms previously quite unknown, yet *natural selection* might act on these individual differences with sufficient rapidity to be in accord with the known facts of evolution; for certainly there is a far longer period of time placed at our disposal, as it were, than was formerly believed to be the case. If the hypothesis here suggested should prove to be true, the accusation of insufficient survival value so often brought against small mutations would fall to the ground.

## (4) The problems to be solved include the appearance of new forms and the bifurcation of species.

It has also often been urged against the theory of natural selection that there are no grounds for anticipating either that varieties far beyond the existing limits of a species would often make their appearance, or that such varieties as did appear would be of a kind possessing qualities making for survival; and perhaps doubts of this kind may have led certain naturalists to devote too much attention to exceptional and infrequent mutations. If such a system of opposing mutations due to *transfers between the allelomorphs* were in operation, as is here suggested, these unaccounted for results are, however, those which might be expected to occur. Imagine a group of freely inter-breeding organisms in which the effects of the centrifugal and centripetal influences had produced a position of stable equilibrium as regards the variation of the group, a condition in which the individual differences were consequently evenly distributed about a mean ; and also imagine that a number of the individuals on one side only of the mean as regards some character or quality were lopped off by natural selection ; then it would obviously follow that the mean of the whole group in regard to that character would be shifted somewhat in the direction in which the selection was acting. In these circumstances the counterplay of the centrifugal and centripetal influences would continually, as it were, keep striving to redistribute the individual differences existing within the group so as to make them centre about this new mean ; with the result that some, but not all, of the lopped off types would be replaced as the result of fresh centrifugal mutations, whilst on the other flank, not curtailed by natural selection, this same kind of mutation would result in organic forms *beyond the previously existing limits* of the group making their appearance. Forms which had been kept out of existence, as it were, by the centripetal influences, would now begin to appear, but only in the direction likely to make for survival. Natural selection might under these hypothetical conditions go on ceaselessly modifying the characters

of a species to an indefinite extent so as to make
them better adapted to its surroundings, the existing
limits of the variety or species forming no impediment
to this progress. No doubt if the mutations were
small in proportion to the individual differences
brought about both by the rearrangement of the
factors and by environmental influences, the process
would be a slow one; but though slow it would be
none the less sure.

Evolution might proceed, and the characteristics
of a species might be changed to an indefinite extent,
in the manner above described, but the total number
of species in existence could not thus have been increased. As to the way in which the multiplication
of species has been brought about, all agree that
isolation may have played an important part; for
natural selection, or indeed any other evolutionary
agency, is not unlikely to have acted differently in
different localities, and thus to have produced a continually increasing divergence between the isolated
groups, until they had come to differ sufficiently to
be described as two distinct species. But if, after
such a differentiation had occurred, the physical
barriers which separated the species had been re-
moved, and if individuals composing them had all
remained, like their ancestors, perfectly fertile *inter-
se*, then the hypothetical centripetal influences would
have begun, by over-mastering the centrifugal mutations, to have slowly reduced the divergence in
characters between the two previously separated
groups; and this would have gone on till the opposing influences had balanced each other, and a
single species had thus again been formed, with no
greater dispersion about its mean than was formerly

the case. If, on the other hand, the two isolated groups had become infertile *inter se* during their isolation, on the removal of the barriers inter-breeding would not have had any tendency to obliterate the differences between them, and they would have remained as distinct species, each, as we actually find in nature, including individuals uniformly distributed about a mean as regards their qualities and each having approximately the same range of variation. The results of the removal of the barriers would, therefore, have entirely depended on whether infertility had or had not been produced by isolation. The important question is not, therefore, whether species can or cannot inter-breed and produce fertile offspring, but rather whether they have *as a fact* freely bred together for long periods of time, offspring of undiminished fertility having actually resulted from these unions. If it can be proved that in many cases two species or varieties have inter-bred quite freely for long periods of time and yet have remained specifically distinct, such a fact would no doubt finally dispose of the mutation hypothesis here suggested for consideration. In the absence of any such proof, may not a species merely be a freely inter-breeding group of individuals, prevented by natural boundaries, by natural infertility or by other impediments, from breeding with other groups, and prevented by the interplay of opposing influences either from becoming all absolutely like each other or from covering more than a certain range of variation in inborn qualities?

It is, however, hardly conceivable that isolation could be made to account for all the innumerable cases in which a species must, as it were, have split in two to form two new species, and it would seem that it must have been possible for bifurcation to have taken place at times without the aid of natural barriers. If we assume that, for some unexplained reason, organisms do become more and more relatively infertile as they come to differ more widely from each other, then it will be seen that bifurcation does become a possibility, granted the existence of these small and frequently occurring opposing mutations. What the effect of the infertility between the individuals composing a species being greater in proportion to the differences between them, or, in other words, of the progeny resulting from unions between individuals resembling each other being especially numerous, would be on the range of variability is doubtful. But though an equilibrium as regards variation would, it may be assumed, be normally established, yet this condition of stability might be upset by the action of the opposing mutations in the following circumstances. Qualities making for survival are not infrequently negatively correlated with each other; as, for example, strength with agility, long round legs making for rapid movement on land with short flat legs if for movement in air or water, conspicuous and attractive colours with invisibility, power of acquiring traditional knowledge with rapid development in youth, etc., etc.; and these qualities might come to be associated in groups by the process already described as making for adaptation. Now if to be relatively well developed *in one or other* of such negatively correlated qualities or groups of qualities was more advantageous to the organism than to be moderately well developed *in both* of them, then natural selection would tend to eliminate the mediocre individuals. A relative reduction in the number of these central types would lessen the number of matings between them and the extreme types, and by reducing the tendency towards uniformity due to such matings, would increase the range of variation of the species under the assumed conditions ; this increase of range would *ex hypothesi* increase the relative infertility of the extreme types ; this increase of infertility would again react on the range ; and so on. Moreover any assortive mating taking place in consequence of the extreme types coming to inhabit different areas, or from any other cause, would also increase the variation. In fact an assumed infertility dependent on the amount of divergence between the individual members of a species, together with a relatively greater survival value possessed by the extreme forms, might destroy the equilibrium between the opposing mutations in such a manner that the only result could be the formation of two stable and completely relatively infertile species out of the extreme types of a group of individuals originally perfectly fertile *inter se*. If we assume that divergence in form often tends to carry with it relative infertility, and also that these opposing mutations do exist, then the bifurcation of species without isolation presents no formidable difficulties.

## (5) Infertility between species and facts connected with pure lines have also to be explained.

The foregoing explanation of bifurcation is, however, dependent on infertility arising when organisms come to differ from each other even to a moderate

extent, and no reasons have yet been given for anticipating that fertility and similarity are likely to become
correlated. Darwin thought that it was impossible
for natural selection, by its action on individuals, to
have increased by slow degrees any incipient sterility,
whether that existing between varieties, or that of
their hybrid offspring; but did he sufficiently consider
the possible results of competition between large
groups of organisms? If the individuals of any
Phylum, or other large division into which organisms
are divided by naturalists, had their method of reproduction so constituted as to result in no infertility
existing between divergent forms, then no species
belonging to that Phylum could split in two in the
manner above described without the aid of isolation
and two new species could not have thus been produced out of one original type. With the aid of
isolation such a bifurcation could no doubt have taken
place; but, if inter-breeding makes for uniformity, the
two species thus formed would only have remained
distinct as long as the natural barriers separating
them continued to be effective, and many species
permanently competing with each other in the same
area could not thus have been created. On the other
hand, if a Phylum were endowed with a method of
reproduction which did result in a certain degree of
infertility arising between organisms even moderately
differentiated by mutation, then even if this infertility
was capricious and irregular, yet many species might
thus be enabled to bifurcate and to produce two new
species, whether aided by isolation or not. The species
composing such a Phylum, by thus increasing in
numbers and differentiating, would in consequence
be better able to seize on all the conflicting oppor-

tunities offered to them by nature and thus to more closely adapt themselves to their surroundings. In the case of the first mentioned Phylum, however, where complete fertility between divergent individuals was retained and where the differentiation of species not separated by barriers so as to suit such conflicting conditions would be impossible, the qualities of the individuals composing each species would necessarily be formed more as the result of a compromise between incompatible demands. The competition between these two groups of organisms, which might have been going on for an inconceivably long period of time, might therefore be compared to that between artisans, each skilled in his own special trade, and jacks-of-all-trades, each trying to play many parts at once as well as may be. Granted that inter-breeding makes for uniformity, and that the hypothetical mutations are really operative, any Phylum fortuitously endowed with the quality of relative infertility naturally arising between diverging types would slowly but surely sweep any Phylum not so endowed off the face of the earth. No doubt if the sterility between the divergent individuals became too great, such a state of things would be injurious by reducing the population and in other ways. It does not on the whole, therefore, appear to be very surprising under the assumed conditions that a gradually increasing infertility should be a common but not universal result of divergence of character.

Since writing the above, Mr R. A. Fisher has suggested another factor tending to produce certain forms of infertility between varieties closely resembling each other, which is more immediate in its effects. In the case of two varieties, each adapted to its own special surroundings, between which inter-breeding takes place, the hybrids thus produced will as a rule be inferior in this respect to both the parent types, and will therefore be reduced in number or exterminated by natural selection. It follows that those individuals which have the least tendency to be fertilized by or to pair with individuals outside the ranks of their own variety will have most descendants; and natural selection will thus tend to foster any natural tendency either to refuse to mate with alien types or for certain forms of infertility to arise between slightly differing organisms. Infertility thus produced would directly facilitate the bifurcation of species, and would also thus in the manner above suggested promote the survival of certain types of organisms when in competition with other types which inter-breed more freely. We cannot, however, thus account for the sterility of hybrids.

It may be worth noting that if the individuals composing a species, or rather some freely inter-breeding group, are widely scattered about their mean, that is if they differ much in structure *inter se*, this state of things would be harmful; for it would prevent as close an average adaptation to their environment as might obtain if there were more similarity between the individual organisms. On the other hand, if the individual members of a species all resembled each other very closely, this state of things would also be injurious; for it would tend to make the action of natural selection very slow. A compromise between these extremes, or a moderately wide range of differentiation between the individuals of a species, would therefore seem on the whole to be most advantageous; and such a moderate differentiation as we do actually find in nature may have been brought about in the manner suggested above to account for infertility having arisen between species, that is as the result of the slow selective effects of competition between great groups of organisms through vast periods of time.

Lastly, in regard to the belief that evolution has in any degree been the result of the selection of small mutations, the arguments against this belief which are based on the results obtained by experiments with 'pure lines' must be considered; for the individual differences existing in pure lines have been found not to be heritable by certain investigators, though not by all. This result may have been due to the selective experiments not having been carried on long enough to detect such an exceedingly slow process as that assumed to be in operation. When experiments have been made with cross-fertilization, selection has often produced marked effects; but the fact that *visible* results have thus been obtained may have been entirely due to the elimination of certain existing types, and not to true mutations. Now may not any differences existing between the gametes when united be a cause or influence promoting mutation? If this should be the case, breeding experiments with pure lines, in which the gametes are assumed to be absolutely identical, either might have to be carried on for a great number of generations in order to secure visible results, or might fail altogether. No doubt if organisms do exist which never conjugate, then for evolution to be possible, mutations must occur at times without the union of any gametes; and if mutations do occur in the case of such asexual organisms, there could be no counteracting tendency towards uniformity produced by inter-breeding, and such organisms could not have been grouped into species in the way here suggested. Do any organisms exist which certainly never conjugate, and, if so, are they less clearly separated out into varieties and species than is the case with organisms endowed with sex? Similarly are there any plants which, without doubt, being always self-fertilized, have for long existed as pure lines; and, if so, are they more or less varied in type than are plants which are cross-fertilized? Here are suggested lines of enquiry which might help to solve this riddle.

It may perhaps here be urged that no explanation has here been given of the way in which, in the course of the evolutionary process, the Mendelian factors in individuals belonging to the higher types of organisms have become more numerous than in their remote ancestors. But until we know what a factor really is, it is not worth while attempting to imagine in detail how new factors may have arisen. It seems more likely, judging by analogy, that genes divide or give birth to other genes rather than that new genes are brought in from outside; and this process of division may also have been stimulated by any dissimilarity existing between the gametes when united. Then again it may be said that we have only been considering quantitative differences between allelomorphs, and not such as are truly qualitative. But can quantity and quality be thus clearly divided? Is it not always possible to express the differences between the characters of any two organisms in terms of the more of one quality and the less of another? If these doubts as to qualitative differences refer to the differences produced by large

mutations or to the large steps on which mutationists
rely in building up their theory of evolution, it need
only here be said that there is no intention to rule
out these large mutations altogether as a factor in
evolution, but merely to indicate that they alone cannot account for many of the adaptations found in
nature. Lastly it will be said, and said truly, that we
have no proof that transfers can take place between
two allelomorphs of such a nature as to affect in
opposite directions the organisms springing from
them; but in the half light of our existing knowledge,
such a supposition, if regarded as no more than a
supposition, may perhaps be worthy of consideration
when groping for the truth.

**(6) In experiments designed to test this
hypothesis natural conditions should be imitated as far as possible.**

If such an hypothesis as that here suggested is
worthy of consideration, what should be the nature
of the experiments instituted to test it? As to the
hypothetical effects of inter-breeding, tall and short
sweet peas might be crossed together for half a cen-
tury or more. By that time the homozygous offspring
should show less difference in stature than was the
case with their ancestors. In regard to experiments
made with the object of directly detecting the existence of the hypothetical small mutations, if they
consisted in the mating of individuals closely resembling each other, then the investigator might be
experimenting under the conditions most likely to
insure that the mutations would be very minute or
infrequent; for this would be the case if the suggestion made with reference to pure lines proved to be

true, namely that these mutations are caused by differences between the uniting gametes. On the other hand if the selected parents were widely dissimilar, then the individual differences between the offspring due to the rearrangement of the allelomorphs would be as a rule proportionately large and proportionately likely to mask any true mutations concurrently appearing. Little surprise should, therefore, be felt if such experiments are slow and difficult.

If we desire to ascertain whether *natural* selection has in truth been operative in *nature* and in what way, should we not *imitate nature* as far as possible in our operations? As a general rule, at all events, we find in every freely inter-breeding group in nature that the measurements of the characters of the organisms are uniformly distributed about a mean, and, if so, the character to be investigated experimentally should be one that is thus distributed. In practice the characters selected for experimentation have often been such "as show some sign of variation," by which is meant that the differentiation is not quite continuous. This may, however, be due to the existence of some unknown impediment to the free formation of the hypothetical small and frequent mutations; and, if so, such a character is one especially to be avoided. Or it might indicate that there had been in the past a mixture of breeds and that inter-breeding had not taken place long enough to obliterate all traces of the mixed origin of the organism in the way in which it is here suggested that such traces would be obliterated in time. In short if we desire to see what natural selection can have effected by operating on uniformly distributed series of organisms, such uniform series should be selected as the ones on which the experiments should be made.

If the mutations on which selection acts may possibly be in large measure the result of differences between the uniting gametes, another point has to be held in view in experimentation. Nearly all experimental breeding is carried on by the selection of the extreme forms on one side of the mean, and by breeding only from them. No doubt in this way apparent results will be most rapidly obtained; but, as already remarked, these results may be entirely due to an increase in the proportion of these extreme types, and not in any degree to any actual mutations of the gametes. No doubt also when breeding only from extreme forms the gametes might never be very widely dissimilar, and in such circumstances it has been assumed that the centrifugal mutations would tend to prevail over the centripetal influences. To get the most rapid evolutionary results, it might, nevertheless be best to copy nature and, after eliminating all extremely undesirable types, to allow the remainder to breed together at random. In this way there would be a wide diversity between the uniting gametes, with the result, it has been suggested, that *both types* of mutations would be greater or more frequent than would have been the case if a more rigid selection had been made. The tendency thus produced for new forms to appear beyond the previously existing range of differentiation might in this way be so much enhanced as to make this method of procedure produce the most rapid results *in the long run* in spite of the countervailing influences above mentioned. Should this be the case, that is to say if extreme types ought to be eliminated rather than selected, if moreover the character selected for study should be one perfectly distributed about a mean, and if the effects produced in each generation are likely to be very minute, what examples are there of experiments conducted on these lines? I know of none. With domestic animals efforts have generally been made to breed only from extreme types; and remarkable results have certainly been thus obtained. But the great differences in structure between certain breeds of dogs, for example, may have been due to the fancier having seized on some sport likely to attract immediate attention; and such modifications can hardly be brought forward in support of a belief in the existence of small mutations. We should rather look to the slow changes, such as those which have resulted in all dogs clinging to human beings or have differentiated the cart horse from some of his smaller relations, if seeking for a counterpart to the evolutionary methods which, it is suggested, may have been in operation in nature.

The points which I wish to emphasize in this pamphlet may be summarized as follows. Recently there has been a growing belief in the efficacy of the inheritance of acquired characters and in the direct effects of environment as factors in evolution, these being agencies on which Darwin relied to some extent. Here it has been maintained that standing alone they cannot account for evolution, even if this tendency to revert to Darwin's views should prove to be thoroughly justifiable. Then as to large mutations, even though it may be right to include them amongst the evolutionary agencies, yet they cannot be relied on to fill the gap left in the explanation of the origin of existing organisms. We must rely in a

measure on small and frequent mutations, or admit
that no explanation is now forthcoming; and certainly
it was on the natural selection of such mutations that
Darwin laid the greatest stress. The difficulty of
proving the existence of small mutations has led
certain experts to deny their existence, a denial which
has had the unintended effect of making many doubt
the whole theory of evolution. But the smaller the
mutation the more difficult does it become to prove
its presence; and to assume that minute mutations
are constantly occurring is a legitimate hypothesis if
it best fits in with all known facts. If we make this
assumption, then it follows that certain influences
making for both uniformity and diversity are con-
tinuously at work within the ranks of every variety.
Without ever ceasing to be operative, these opposing
influences may balance each other at some one point;
and when such a position of equilibrium has been
reached in any freely inter-breeding group of organisms, a definite range of variability of moderate
dimensions in their hereditary characters will thence-
forth be permanently maintained. Here is a basis on
which natural selection can act, and if so acting it
could not be said that small mutations had insufficient
survival value. The main difficulties in the theory of
organic evolution are how to explain the uniformity
of useless characters, the adaptation of structure to
environment, and the origination of specific differences; and it seems that all these difficulties are to
a large extent overcome if the hypothesis of small
and frequent mutations may be accepted. As to the
formation of species, we should expect any freely
inter-breeding group to remain unaltered for an indefinite period of time under certain conditions; for

if the mediocre forms did fit their environment better than those at either end of the series, no evolutionary change could be brought about by natural selection, though the normal degree of differentiation would be continually maintained by the opposing influences. Taking the imaginary case of a perfect gradation of environments having resulted in a perfectly graded series of organisms, all suited to their different environments, the assumed conditions would almost certainly result in such a series being in a state of unstable equilibrium. The population would be certain to be more numerous in some places than in others; and, granted that inter-breeding does make for uniformity, the more numerous the group the less would be the injury done to it by an admixture of foreign strains, and the more likely would it be to prevail over its neighbours. A perfectly graded series would break up into a number of distinct varieties without intermediate forms between them; and as it has been seen that a tendency might arise for infertility to become correlated with individual differences, a continuation of this process, if the relative infertility became considerable, would make the differences between these varieties become sufficiently marked to allow them to be promoted to the more dignified rank of species.

There is perhaps little novel in the above suggestions, and they will, I believe, be largely endorsed by many experts, a class to which I do not pretend to belong. I have, however, also suggested that the union of differing gametes may cause a peculiar kind of mutation by leading to something in the nature of transfers taking place between the allelomorphs; a suggestion which may not be welcomed by many experts. It does, however, seem to me to help to explain the difficulty of detecting mutations in pure lines, the absence of abrupt variations in freely inter-breeding groups, the general similarity in the range of variation exhibited by closely allied but completely separated varieties, and the continuous and unlimited appearance of new forms in the direction in which selection is acting. If this suggestion should be ruled out by experts as being unnecessary, no harm will, at all events, have been done by making it. What I have mainly wished to indicate is that an hypothesis can be framed without much difficulty which does fit with the facts, provided the occurrence of small and frequent mutations be admitted. No doubt the bright light of Mendelism has modified or destroyed certain arguments relied on by Darwin, in whose time that light was not shining. But if we are bound to rely on small and frequent mutations to a considerable extent in explaining organic evolution, and if there is no bar to our so doing dependent on facts, then the edifice erected by the author of the *Origin of Species* still remains intact in all its essential features.